volume 4 DEVELOPMENT AND UNDERDEVELOPMENT 1945–1975	volume 5 THE GLOBAL COMMUNITY 1975–2000	volume 6 INTO THE 21ST CENTURY 2000–	chapter	topic
THE END OF PEASANT CIVILIZATION IN THE WESTERN WORLD	INDUSTRIALIZED AND MULTINATIONAL AGRICULTURE	THE PROBLEM OF SURVIVAL AND THE PROMISE OF TECHNOLOGY	1	Land, agriculture, and nutrition
OLD DISEASES IN THE THIRD WORLD AND NEW ACHIEVEMENTS IN MEDICINE	OVERPOPULATION, DEMOGRAPHIC DECLINE, AND NEW DISEASES	POPULATION, MEDICINE, AND ENVIRONMENT: A RATIONAL UTOPIA	2	Hygiene, medicine, and population
THE AGE OF PRIVACY: HOUSING, CONSUMER GOODS, COMFORT	MEGALOPOLISES IN THE THIRD WORLD; MULTIETHNICITY IN THE WEST	TOO MUCH AND TOO LITTLE: THE MISERY OF WEALTH	3	Living: environment and conditions
WHITE-COLLAR WORKERS, MANAGERS, AND LABOR PROTEST	AUTOMATION AND DECENTRALIZATION IN THE POST-FORD ERA	"THE END OF WORK" AND THE NEW SLAVERIES	4	Labor and production
NUCLEAR ENERGY: THE GREAT FEAR, THE UNCERTAIN HOPE	THE SEARCH FOR ALTERNATIVE ENERGY SOURCES	NEW FRONTIERS IN ENERGY	5	Raw materials and energy
THE PRODUCTION OF THE AUTOMOBILE	ELECTRONICS AND INFORMATION SCIENCE	WORK WITHOUT WALLS	6	Working: environment and conditions
MAN AND GOODS ON FOUR WHEELS	THE AIRPLANE IN MASS SOCIETY	FROM THE EARTH TO THE COSMOS: THE EXPLORATION OF SPACE	7	Transportation
EVERYDAY ENCHANTMENT: THE TELEVISION	THE INFORMATION AGE: COMPUTERS AND CELL PHONES	CYBERSPACE: THE WEB OF WEBS	8	Communication
PARALLEL ROADS: DEVELOPMENT AND UNDERDEVELOPMENT	THE COLLAPSE OF SOCIALISM AND THE RISE OF NEO-CONSERVATISM	THE MANY FACES OF GLOBALIZATION: LOCAL WARFARE AND THE GLOBAL COMMUNITY	9	Economics and politics
MOVEMENTS OF LIBERATION AND PROTEST: THE THIRD WORLD AND THE WEST	FEMINISM, ENVIRONMENTALISM, AND THE CULTURE OF UNIQUENESS	UNIVERSALISM AND FUNDAMENTALISM: THE NEW ABSOLUTISM	10	Social and political movements
CONSUMERISM AND CRITICISM OF THE CONSUMER SOCIETY	THE INDIVIDUAL AND THE COLLECTIVE	AFTER THE MODERN: ENVIRONMENTALISM, PACIFISM, AND BIOETHICS	11	Attitudes and cultures

THE ROAD TO GLOBALIZATION
Technology and Society Since 1800

In private and in public, at work or at play, in every stage of life, we live with technology. It becomes ever more present, and our perception of its artificiality fades through daily use. Within a very short time of their emergence, new possibilities seem to have been with us always, and the new almost immediately becomes indispensable. The choices that technology dictates and the paths that these choices take appear to be the only choices and paths possible—undeniable, unquestionable—and we perceive as natural the constructed world in which we live.

Despite the opportunities that technology affords us, and the promises that it makes constantly, we greet it with a general discomfort, an uneasiness that often does not reach the conscious level. But the manifestations of environmental crises can no longer be considered in isolation. The Westernization of the world marches in step with the widening—and already yawning—chasm between north and south, as well as with the emergence of aggressive localism. War seems to have resumed its role as a common tool in international confrontation. New diseases alarmingly outpace scientific discoveries, and biotechnologies and genetic experiments obscure the line between the human and the inhuman. The importance of the question of meaning has not been lessened by the decline of the sacred; but this question seems to find no place in the universal logic of growth that overcomes difference to guide governing bodies as representatives of economic and financial power.

A renewed uncritical faith in Progress on one hand and a demonization of "techno-science" on the other are often associated with a lack of context that comes of technology and with the dominance of a logic that neglects history. This logic can verify correctness in predetermined ways, but it does not comprehend the complexity of the greater process of change: it appreciates the present and the immediate future, but it cannot perceive itself as part of a larger historical evolution.

The first aim of a social and cultural history of the technology of the last two centuries, then, is to offer a careful and coherent study of the roads that have led to the development of modern culture. The basic objective of this series is the reconsideration of the innovative changes that have taken place and their diffusion over time, rather than a description of their first appearances. These innovative changes have marked and continue to determine our daily lives, the way we work, our relationships, and the points of view that contribute to global diversity.

It is important to recognize that our interpretations of the 19th and 20th centuries are centered on the men and women of the West, on their histories and cultures. This is undoubtedly a biased point of view, and it would be misguided to think that this partiality could be overcome by a simple updating of knowledge. The changing of a point of view that is rooted in history probably requires insight into processes that operate well beyond our perception. Perhaps the globalization that is underway, with its various worldwide effects, is establishing itself through precisely this mechanism: it is forcing a confrontation among lifestyles and different cultural models in new, absolute terms.

In general, the common historiography treats technological innovations only in brief digressions, glossaries, or chronologies of inventions and inventors; but we cannot fill in its gaps by constructing a separate history. Our realization of the economic, social, and cultural importance of industrialization, and our perception of the process as uninterrupted and ever more pervasive, have caused us to re-evaluate both the transformation itself and the new landscapes that industry has created—linking technology to economics and to politics, and systems of labor and production to culture and to social movements.

We can group as the Age of Technology the events that have been paving the road to the future for the last two centuries. Understanding the risks and the opportunities involved in so rapid a transformation of our world will require a change of mind and an updating of our culture—both of which are impossible without a broadening of knowledge and a renewal of historical consciousness.

1
THE INDUSTRIAL REVOLUTION
1800–1850

PIER PAOLO POGGIO
AND
CARLO SIMONI

ILLUSTRATED BY GIORGIO BACCHIN

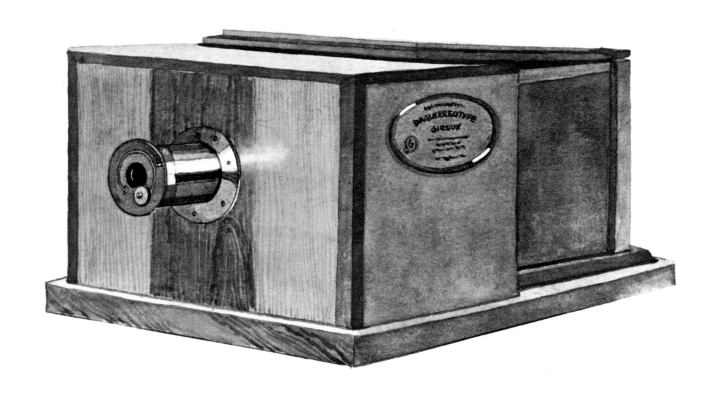

English-language edition for North America © 2003 by Chelsea House Publishers.
All rights reserved.

All rights reserved. No part of this publication may be reproduced or transmitted in
any form or by any means without the written permission of the publisher.

Chelsea House Publishers
1974 Sproul Road, Suite 400
Broomall, PA 19008-0914
www.chelseahouse.com

Previous page: Louis Daguerre's photographic machine, built in Paris in 1839 by Alphonse Giroux. (Drawing by Paola Borgonzoni.)

Popular French print of the time of the revolutions of 1848, entitled Un marché sous la République Universelle et Sociale (A Market Under the Universal Socialist Republic).

Library of Congress Cataloging-in-Publication Data

Poggio, Pier Paolo.
The Road to globalization : technology and society since 1800 / text by Pier Paolo Poggio, Carlo Simoni ;
illustrated by Giorgio Bacchin.
p. cm.
Includes index.
ISBN 0-7910-7092-1
1. Europe—Social conditions—Juvenile literature. [1. Europe—Social conditions—19th century.
2. Europe—History—1789-1900.] I. Simoni, Carlo. II. Bacchin, Giorgio, ill. III. Title.
HN373 .P64 2002
2002007848

© 2001 Editoriale Jaca Book spa, Milan
All rights reserved.

Original English translation by Karen D. Antonelli, Ph.D.

Printing and binding by
EuroLitho spa, Cesano Boscone, Milan, Italy

First Printing
1 3 5 7 9 8 6 4 2

INTRODUCTION

The first volume of this series is dedicated to the Industrial Revolution, a sequence of events and developments that provided the background for everything that followed—of change so great that its extent can be compared only to the revolutionary invention of agriculture in the Neolithic Age. The countryside, the cities, the peoples, the economy, the political ideas, the opinions, and every other aspect of life changed as a result of widespread technological innovations.

Industrialization, which had begun quietly in the British Isles with the introduction of the mechanized loom and the steam engine, spread to other European countries in the first decades of the 19th century. Simultaneously, major ideologies came into play that would confront each other in a variety of ways over the next two centuries: liberalism, nationalism, and socialism.

The landscape of cities and regions changed with the diffusion of the factory, the railroad, gas lighting, the telegraph, and the steamship. Goods and people traveled, and with them traveled ideas and culture—especially across the Atlantic Ocean toward America, where these new views would find fertile ground in which to take root.

Tompkins H. Matteson, The Last of the Race *(1847), New York Historical Society. Matteson depicts the extinction of the American Indian tribes, a consequence of the "taming of the West".*

LAND, AGRICULTURE, AND NUTRITION

1. A Revolution in Agriculture and the Last Famines in Europe

In order to increase agricultural production, a new method of cultivating land, based on the periodic rotation of crops, was tested in northern Italy in the 16th century. Nearly two centuries later, drawings appeared in France of new machines capable of helping farmers with their work. Although significant agricultural innovations became established in the 18th century, the mechanization of agriculture would take hold only in the second half of the 19th century. Before coming into common use, discoveries, innovations, and technological improvements have always had to wait for a social atmosphere conducive to their acceptance. This agricultural revolution, begun in England in the 18th century, spread only to places where great landholders, not limiting themselves to living on the revenue from their properties, were willing to invest capital in their lands.

These developments occurred in countries in which land could easily be bought and sold and where it was already considered a means of amassing new wealth. On the other hand, in the regions in which traces of feudal culture remained, such as Russia and southern Italy, these new techniques remained unknown for a long time. For the same reasons, the newly designed farming machinery that was coming into use in the United States would spread only slowly in Europe, throughout the 19th century.

In the middle of the 19th century, notwithstanding increases in productivity, European agriculture remained subject to famines. These resulted from several cold and wet summers, such as

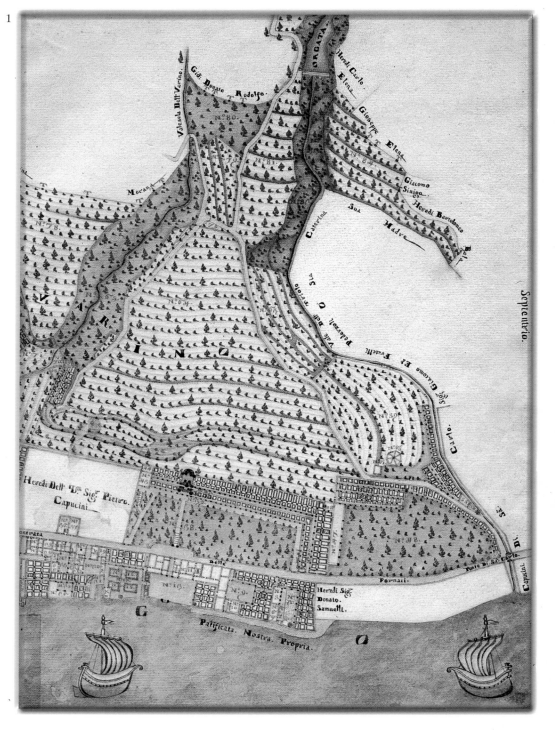

1. An 18th-century map of an area under cultivation with lemons, olives, and grape vines on Lake Garda in northern Italy. As the neat order of the agrarian landscape indicates, considerable investment and expanded technology were employed in agricultural circles beginning in the 18th century, both on the great land holdings of the plains of the Po Valley in northern Italy, and in certain areas dedicated to specialized crops.
2. A mechanical seeder from an illustrated plate in Diderot and D'Alembert's Encyclopédie. *Between the end of the 18th century and the first decades of the 19th, the seeder, like the plough before it, was studied carefully to improve its performance and increase the production of the lands. The mechanization of agriculture led to a greater need for iron and thereby affected the development of the mechanical and steel industries.*

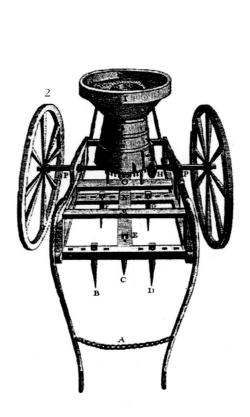

those of 1816 and 1817, or from almost exclusive dependence on a single staple crop. This was the case in Ireland, where the potato was infected by a fungus that destroyed the crops in 1846 and 1847, causing a staggering decrease in population.

3. Vincent Van Gogh, The Potato Eaters. *When the great Dutch artist painted this canvas in 1885, the poverty of farmers in Europe could be represented by a family whose only sustenance was potatoes.*
4. The Sahel. *For a great part of the world, famine is not just a memory; even at the end of the 20th century much of Africa's Sahel region, south of the Sahara desert, was subject to famine. This photograph by E. Turri was taken in Mali.*
5. Jeff Katz, The Potato Famine, ca. 1846. *The 1846–1847 famine in Ireland, here shown in a depiction of squalor and desperate misery, hit a population that had relied on the potato as its basic food source since the 17th century. Thanks to that staple, Ireland had reached a population level of approximately 8 million people. In less than 10 years, the Famine had resulted in one and a half million deaths, and the same number of Irish men and women were forced to abandon their country. (Photograph FPG International/Marka.)*

HYGIENE, MEDICINE, AND POPULATION
2. The Demographic Revolution

A demographic equilibrium had existed within agricultural societies: the high number of births had been compensated for by an equally high number of deaths. The limitations to increases in population were due consistently to several factors. First, the production permitted by the agricultural technology in use at the time was severely limited by the restrictions imposed by feudal lords, who were unwilling to finance investments that might increase or even maintain the production of their lands. Thus the population, after a period of increase, would enter a period of progressive debilitation and thus become vulnerable to the epidemics that continued to spread. (The bubonic plague, which had been responsible for waves of mortality that in some cases had reduced the population by half, had essentially disappeared from western Europe by the end of the 17th century.)

Some scholars of historical demography claim that the increases in population that took place in the first decades of the 18th century cannot be explained by either the decrease in the mortality rate or new health measures like the one adopted by Marseilles, France in 1720–1721—a sanitary cordon to keep the latest epidemic within the city limits. Important innovations in the field of medicine—for example, the vaccine against smallpox, another terrible disease—took root only in the following century. This is also true of the use of antiseptics and the establishment of a network of hospitals. Rather, the determining factors in the increase in European population can be found in improvements in agricultural output and in the increased availability of food.

The reduction of famine, almost its complete elimination, was caused by the trading of agricultural practices and information among agrarian cultures. This trading resulted in the widespread consumption of the potato in Ireland and Germany and of maize (corn) in northern Italy. Beginning with the

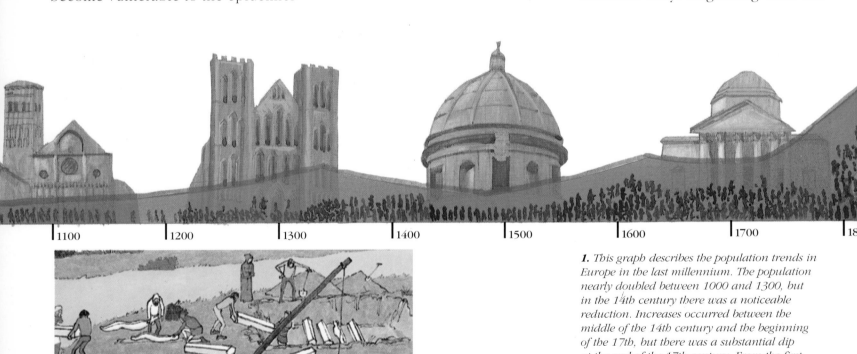

1. This graph describes the population trends in Europe in the last millennium. The population nearly doubled between 1000 and 1300, but in the 14th century there was a noticeable reduction. Increases occurred between the middle of the 14th century and the beginning of the 17th, but there was a substantial dip at the end of the 17th century. From the first quarter of the 18th century onward, there were no further interruptions in the increase, which became much faster.

2. The plagues were a scourge for Europe. Scenes of mass burials began to appear in the Middle Ages and continued through the 17th century. (Drawing by S. Corsi.)

3. Like the plagues, wars decimated the population. Wars were fought throughout the year, calling men away from the countryside and thereby causing a cyclical impoverishment. (Drawing by A. Molino.)

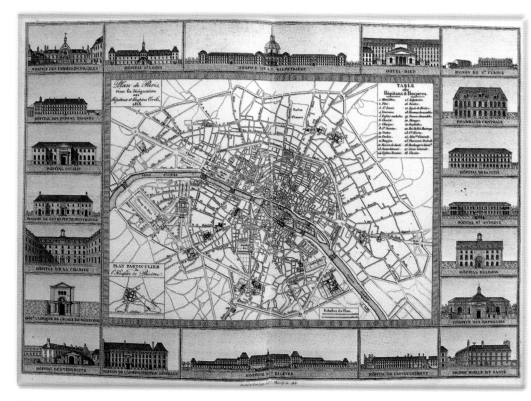

4. *A map from 1818 identifying the hospitals of Paris. In the 19th century, hospitals slowly began to expand beyond their function of caring for the urban poor. Although caregiving continued, hospitals also became centers of research and experimentation.*

5. *Edward Jenner (1749–1823), an English local medical officer, was the first to experiment on a child infected with smallpox. He inoculated the boy with the pus of bovine smallpox, which was considered less aggressive than the human kind.*

6. *The potato and corn were two primary sources of European popular nutrition in the 19th century. (Drawing by L. Pieri.)*

second half of the 19th century, the development of new modes of transportation and of an international agricultural market made the agricultural crises that had previously struck the continent just a memory.

LIVING: ENVIRONMENT AND CONDITIONS

3. Cities, Neighborhoods, and Villages: Middle-Class and Working-Class Homes

1 In the 19th century, masses of landless farmers headed toward the cities of western Europe. This was a result of the establishment of capitalist controls in the countryside and a vast increase in population. This migration led to new models for the organization of space and new ways of life within the cities.

The traditional mixture of social classes became less common. Ever more frequently, working-spaces were separated living-spaces. Similarly, in middle-class families a division arose:

1. The village of New Lanark in 1825. In this English colonial center, inaugurated on January 1, 1800, Robert Owen performed a social and economic experiment that no one had tried before: he invested a part of the company's profits to improve the living conditions of the workers. In addition to houses for the workers and a public kitchen, Owen established a "School for the Formation of Character." Manufacturing continued in New Lanark until 1968.
2. This reworked etching by Gustave Doré, which he published in 1872 after a stay in London, shows a dramatic scene of a "slum." In the background, we see the large mercantile port of London, surrounded by factories—a tangible sign of the English dominance of commercial trade in the 19th century.

men became responsible for production, business, and public life, and women were responsible for reproduction, the care of the home and the children, and private life. Even in apartments, distinctions between types of room became apparent: the kitchen and the dining room; the living room and the study; the parents' bedroom and the children's bedroom.

Developments in urban transportation enabled the division of cities according to function: the city center accommodated large shops, financial offices, and the homes of wealthier citizens. The industrial suburbs contained the factories and the homes of the workers. Many of the homes in the industrial zones were poorly constructed. Often they consisted of unhealthful hovels in which mingling of the sexes was inevitable, household furnishings were scarce, and indoor bathrooms were nonexistent.

The workers who lived in "company towns" had better living conditions. These villages, which began to spring up in the 18th century, were erected in mining areas or near the water sources that furnished energy to industry. The experiment of Robert Owen at New Lanark excepted, in the many factory towns that flourished throughout the century across Europe, industrial patriarchy exercised its power over worker's and his family's time and options. The threat of being fired became associated with a notice of eviction from the house that the worker had been allowed to rent.

3. Edgar Degas, The Bellelli Family, *ca. 1860, Musée d'Orsay, Paris. This painting shows the dominant role of the woman in middle-class family life. The home is her realm, and the daughters grow in close contact with their mother. The father, who has his own room, the "study," seems to be a background presence; his world is external, the world of business and social relationships.*

LABOR AND PRODUCTION

4. THE FACTORY SYSTEM: TEXTILES

1. Vincent Van Gogh, detail from Weaver with Loom, *1884. For a long time, textile work had been done only in the home. Unlike spinning, loom work was done by men. In fact, it was the weavers who owned their own machines and had specific abilities who spurred the revolts of the Luddites against the spread of mechanization and the concentration of work in factories.*

2. A water mill in a painting by Claude Lorrain from the middle of the 17th century. The water wheel was the first machine invented in antiquity but came into use in Europe only in the 11th and 12th centuries. First used for the grinding of grains, it was soon used in mines, in paper mills, and in sawmills.

3. For decades, the water wheel also powered textile factories. Until the advent of the steam engine, it was the most efficient motor. A complex system of gears, a long shaft that crossed the entire building, and a series of belts and pulleys allowed the operation of the machines, which were placed on various floors of the building. (Drawing by D. Spedaliere, from a contemporary print.)

At the beginning of the 19th century, commercial use of raw cotton from English colonies enabled the English textile industry to grow to a level thirty times higher than its level of fifty years before. Still, less than 40% of the working public was employed in the agricultural sector. The population had grown in absolute numbers, owing to a general demographic increase. Within a few decades, masses of people, mainly farmers who had been forced to abandon their fields, had boosted the supply of wage labor as well as the demand for inexpensive textiles.

Historians have shown that many factors of the Industrial Revolution had already been operating in England in the 17th century. These factors increased in influence in the last decades of the 18th century and the first decades of the 19th, and then they spread to the rest of Europe. Among these factors, very important at the time, are the availability of mining resources and the accumulation of commercial and agricultural capital. Also, England's international role had a decisive influence; England had earned this through the naval strength that guaranteed its domination of the sea and its resultant domination of the importation of inexpensive products from its colonies and America. Other determining factors were the revolution in agriculture and the establishment of a national market that had no peer in any other country.

At the beginning of the 19th century, most English factories had fewer than 100 workers. Employment was not limited to artisans with machines. Instead, these factories used the method of division of labor that the economist Adam Smith had outlined for the use of textile machines. Textile machines were powered by hydraulic energy at the time but soon thereafter converted to steam power; artisans had been perfecting these machines since

the second half of the 18th century.

But the innovations introduced in factories were not only technological. The factory established a new system of production, one that did not require specific abilities of its workforce. The workers needed only to be able to oversee machines that were inexhaustible—machines that functioned regardless of working conditions or the passage of time. Strict discipline, low wages, and the threat of firing forced men, women, and children to accept, under industrialization, more dangerous and degrading labor conditions than had previously existed.

4. The Arkwright cotton mill in Cromford, Derbyshire, England at the end of the 18th century. The textile factory represented a new presence in the landscape. Because of its great size, comparable only to that of cathedrals, it often appeared menacing to the people of the time.

5. Children at work in a textile factory. Forced child labor was widely used because the children were docile, they commanded much lower wages, and their small hands were able to reattach the threads that the machines frequently broke. This exploitation, which continued in certain areas of production and services into the 20th century, is far from being abandoned today, and it is even on the rise in the southern part of the world and among the immigrant populations of the North.

RAW MATERIALS AND ENERGY
5. Coal, Steam, and Gas

The increasing production of charcoal for iron furnaces had caused deforestation in the English countryside. The need for fossil fuels (like coal) became obvious in the 17th century; these were plentiful in England. The invention of the steam engine, which proved to be extremely versatile, enabled the exploitation of coal deposits.

The steam engine made it possible to drain water from the mines, enabling deeper wells to be dug and thus the production of more coal. This increased supply of coal led to a decrease in its cost, so coal was then used to run the same steam engines that had drained the mines. Thus began the phenomenon of interdependence that characterizes the Industrial Revolution as a type of chain reaction, a self-propelling process.

Greater quantities of coal were also being demanded by the steel industry, which produced an ever-increasing amount of the steel that was used to meet the demand for agricultural and textile machinery.

Using coke instead of coal made it possible to avoid the pollution of cast iron that resulted from coal's sulfuric content. The use of coke also enabled the creation of ovens more productive than those fueled by charcoal; charcoal ovens required greater aeration. Steam provided the solution to all these situations, permitting the use of powerful blowing machines. The increased iron production meant even more raw material for the construction of steam engines—another example of the interdependence between innovation and production.

Even coal became adaptable to other uses. Not only was it a fundamental source of energy, but also, through the distillation process, it became the raw material used to make tar. Tar was later to provide the basis of organic chemistry, and after further processing it became an essential component in the manufacturing of waterproof textiles.

From the beginning of the 19th century, in fact, coal demonstrated an unexpected potential. The first attempts to produce light from gas were made in France and England during this period. This new method of illumination was used primarily in factories and along roads. The use of coal gas to light homes began only after complete combustion could be assured—avoiding any poisonous effects. Starting in the middle of the century, gas was used to heat homes, and after 1870 it was used to cook food.

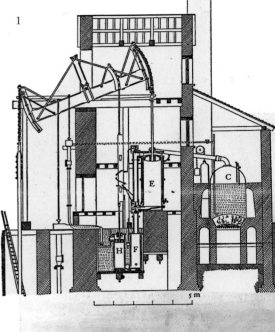

1. A diagram of the steam engine, which was invented and refined in the second half of the 18th century by the Scottish mechanic James Watt (1736–1819). As a product of the early Industrial Revolution, the steam engine can be considered the result of the meeting between traditional technical knowledge and the application of science to researching productive innovations.
2. A landscape of smokestacks and a dense haze of smoke from a German coke factory from the middle of the 19th century. Coke, an excellent combustible fuel derived from coal, appears to have been produced and used for the first time by English brewers. It was used a few years later, in 1781, by Abraham Darby in his blast or smelting furnaces. Darby's foundry produced the cast iron used to construct the first iron bridge, for which the English area of Ironbridge, Shropshire later was named.

3. The steam engine, shown here in an English painting (ca. 1820), was used primarily to increase the efficiency of the pumps that drained the water from mines.

WORKING: ENVIRONMENT AND CONDITIONS

6. THE MINES: WEALTH, TOIL, AND DEATH

Because there was a much greater need for iron and coal during the Industrial Revolution, the work of the miners in the 19th and 20th centuries was notorious for its high degree of difficulty and danger.

Picks and levers, sledgehammers, drills, and explosives were the essential tools of the miner. In the second half of the 19th century, perforating machines appeared in the iron mines and mineral-cutting machines in the coal mines. Before the arrival of steam-driven elevator hoists, the miners reached the bottom of the mines by ladders that sometimes were more than 650 feet long.

An 1840 investigation of English mining practices revealed that the transportation of the minerals was often performed by boys and girls between the ages of 5 and 7. These children were forced to work 12 to 16 hours each day pushing trolleys through underground tunnels. Women dragged carts or carried large baskets filled with coal up long ladders to the surface.

In 1842, the Mines Act kept children under the age of 10 and women out of English mines. In other situations, however, like the sulfur mines in Sicily, child labor continued to play a pivotal role in mine operations.

1. *Scottish coal haulers at the beginning of the 19th century. Before the installation of powerful elevator hoists, the transportation of minerals to the surface required the use of staircases hundreds or thousands of feet in length.*

2. *Early in the 19th century, many European mines, including those of Saxony, Hungary, and the Rhineland, used stepping, the process of carving "steps" from the sides of the tunnels to provide the miners with a working surface.*

The steam engine permitted the use of drainage pumps in the tunnels and of powerful hoisting machines. Mining was always accompanied by the threats of collapsing tunnels and, especially in the coal mines, of a fire-damp—a mixture of methane and air that exploded when it came into contact with the flames of the miners' oil lamps. After a series of tragic explosions in

1812 and 1813, the chemist Humphrey Davy discovered that an iron mesh placed around the open flames of the lamps diminished their heat enough to impede or to prevent the ignition of the explosive pockets of gas. When exposed to this gas, the flames behind their protective grids would lengthen and change in color to a bluish hue. The installation of steam-powered air circulators in 1830 provided another defense against fire-damp, but even after that innovation, this danger claimed victims among the miners for years.

3. A miner at the bottom of a mine shaft, waiting for the trolleys full of minerals to be transported to the surface.

4. Putters, young boys responsible for moving the trolleys loaded with coal through the tunnels. The consequences of this highly unhealthful and laborious work were serious respiratory disease, stunted growth, and a very low life expectancy.

5. Between the two World Wars, manual labor, here depicted in a mine in the Alps near Bergamo, Lombardy, Italy around 1930, continued to play a vital role in the movement of minerals in tunnels.

TRANSPORTATION

7. THE LOCOMOTIVE

In 1804, Richard Trevithick, an English mine owner, came up with the idea of running a steam engine along the tracks used to carry mined minerals. Ten years later, employing the same system, George Stephenson, a stoker and steam-engine repairman, developed the idea of connecting the mine at Killingworth Colliery, located in County Durham, England, with its port of loading.

Like the steam engine, the locomotive resulted from the intersection of different needs. The various means devised to satisfy these needs led to the exchange of ideas and methods. The locomotive facilitated transportation and thus increased the production of coal, the main material needed for its own functioning; this in turn was useful for the production of iron, also necessary to locomotive construction.

Because the train's operation required the construction of railroad tracks, tunnels, viaducts, stations, and telegraph lines, it quickly became one of the principal catalysts of industrialization.

The train represented the most revolutionary innovation of the 19th century. It was capable of modifying the lives of a greater number of people than the factory affected. It also was the primary force in the imposition of a uniform system of keeping time in the countryside. This was essential to not only the efficiency but also the safety of the new means of transportation, as a uniformity of time ensured that no two trains ran on the same track at once. Incidents of this kind did occur, however, providing grist for the mill of opposition to the railway system.

The American writer Henry David Thoreau considered the railroad the profaner of uncontaminated territories, and the Englishman John Ruskin said it was responsible for reducing passengers to human cargo. Still, despite the belief that the railroad would destroy the idea of space—a belief common at the time—many, like authors Heinrich Heine and Hans Christian Andersen, were fascinated by train rides, which allowed the latter to read or study geographic charts while, according to his own description, "the carriage slipped like a sleigh over flat fields of snow" and the locomotive panted like a "wild horse".* Even those who complained that the train did not permit the eye to fix on anything, and thus deprived the traveler of a substantial part of the pleasure of traveling, had to accept, within a few years, the idea that the train was a new way to view the countryside—in its entirety and not merely as a rapidly changing series of objects.

*Quoted from Rinaldi's *Il Bazar di un Poeta*, Biblioteca del Vascello, Rome, 1991; *Il Bazar* is Rinaldi's translation of Andersen's *Ein Digters Bazar* (Gyldendal, Copenhagen, 1842). The translation into English is ours.

1. With Robert Stephenson (the son of the George Stephenson of Killingworth Colliery) as its engineer, the locomotive The Rocket *won the 1829 engine-driven race. The race participants were four vehicles on tracks and a horse-drawn carriage. The Rocket reached the speed of about 28 miles per hour.*
2. Claude Monet, St. Lazare Station, *1877, Musée d'Orsay, Paris. In the urban landscape of the second half of the 19th century, the train station represents a privileged place. A sign of the future, it had a monumental presence that the Impressionist painters did not ignore.*
3. J.C. Bourne, The Tunnel at Bath, *a frontispiece illustration taken from Bourne's* History and Description of the Great Western Railway. *The train's difficulty in mounting steep grades led to the construction of colossal viaducts and very long tunnels. These represented a spectacular proving-ground for 19th-century engineering.*

20

4. The train was a determining factor in the conquest of the American West. In the collective imagination, it became the symbol of a progress that, along with a spirit of adventure and a faith in the power of technology, became deeply meaningful. (Drawing by R. Simoni.)

5. George Inness, The Lackawanna Valley, 1856. The railroad tracks that divided the land and vanished into the horizon, the pinnacle of smoke, and the whistle of the locomotive: the eruption of the train in the tranquil country landscape became, in the 19th century, a constant reference in literature and painting. The train went on to inspire innumerable film sequences. (Photograph from the National Gallery of Art, Washington, DC.)

6 and 7. A photograph taken in a first-class cabin in France around 1900; a third-class cabin in an 1860 sketch by Gustave Doré. The train, like the ship, with its divisions into classes, reflected social dissimilarities and differences in behavior. On the one hand, the habit of reading during a train trip and the tendency to make the train compartment into a private space both were encouraged by a decor resembling the a domestic living room. On the other hand, for the chatting third-class passengers, who at times broke into spontaneous song, the railway car, which was not divided into compartments and was furnished only with wooden benches, offered a space for socializing.

COMMUNICATION

8. The Newspaper and the Telegraph

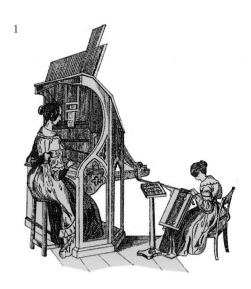

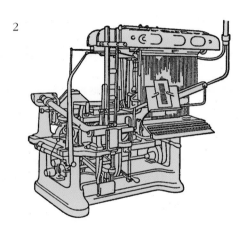

In 1780, the printing of news-sheets surpassed 9 million copies in England. Fifty years later, the number had risen to 30 million. Still, if not for innovations that had accelerated production, guaranteeing timely and widespread diffusion of information not only in England but in the United States and France as well, the newspaper never would have become the most powerful force in forming public opinion in the early 19th century.

It was at the beginning of the 19th century, four hundred years after its invention, that printing saw mechanization. In July of 1814, London's *The Times* issued its first edition printed using a new machine equipped with printing rolls. Thirty years later in Philadelphia, an enormous rotary printing press was in use.

The use of the new machine spread rapidly, even in Europe. Just as improvements in weaving equipment had forced improvements in the spinning machines in the textile industry, so the mechanization of typesetting was required by the printing industry. This came with the invention of the Linotype, which was used for the first time by *The New York Tribune* in 1886. The use of continuous-paper machines in the paper mills led to an increase in production and a decrease in the cost of paper.

The other raw material for the newspapers, obviously no less essential, was news. The flow of news was guaranteed by reporters and press agencies, but their work would not have been possible without the invention of the telegraph. After long trials, the new electric telegraph replaced the optical telegraph, which had been in use for about fifty years, since the time of the French Revolution. The new

machine answered the demand for rapid communication that came not only from the sphere of information-gathering, but also from the railroads, financial enterprises, and organizations of international politics.

From the middle of the 19th century, a thick net of transmissions, at first only by land but later by sea as well, made it possible to disseminate news among the continents with a speed never before known. For the first time, a new development arose that has characterized the contemporary world ever since: a simultaneity of experience.

1. The appearance of the typesetting machine in the first half of the 19th century was the most important development in the history of printing since the industry's establishment.
2. The name "Linotype," which appeared in 1886, was derived from the English expression "lyne o' type"—a line of characters or symbols. A single operator, instead of the three that had been necessary before, could compose a line of type, and the composition itself could be accomplished faster than ever.
3. By 1861, Richard Hoe had developed his enormous rotary printing press, with ten automatic feeders, which allowed the printing of about 20,000 copies per hour. Its operation required the work of 25 men and boys.

4. The telegraph, capable of using the famous code of dots and dashes that was named for its inventor, Samuel Morse (1791–1872). (Photograph by A. Stabin.)
5. The telegraphic cable destined to be placed across the Atlantic Ocean is shown here in the hold of the ship Great Eastern (1865). The cable was almost 3,000 miles in length.
6. The placing of the telegraphic cable in the English Channel in 1850 was performed by the tugboat Goliath.
7. The telegraph operator, an irreplaceable presence in the world of the 19th-century railroad. The telegraph embodied efficiency and security in communication. The long row of poles, running parallel to the track, that supported the wires became symbolic of the installation of the train.

ECONOMICS AND POLITICS

9. THE WEALTH OF NATIONS AND NATIONAL STATES

The nation-state is perhaps the most significant political invention in Europe. It was formed by a long historical process that began with a crisis in the universalistic medieval powers—empire and the Catholic Church—and the affirmation of monarchic absolutism.

Between 1770 and 1790, the formation of the nation-state was accelerated, and for the first time the existence of such a political entity was proclaimed in the name of a sovereign people. This extraordinary newness had been put to the test by European emigrants to North America, where the Colonies decided to join together into a nation, declaring their independence in 1776 and creating the Constitution of the United States of America in 1787.

Two years later, France, the oldest and most deeply rooted European nation-state, re-created itself in the name of popular sovereignty and human rights. The French Revolution began a cycle—one that continues today—in which successive political movements have adopted the principle of national sovereignty.

Thus the model of the European nation-state established itself throughout the world, concurrently with the Industrial Revolution and the rise of industrialization. Because of its industrial primacy, England also acquired a political hegemony, allowing the defeat of opposing nations through trading ability rather than military strength.

When the Industrial Revolution gathered strength, the free market became the central institution of the new economy. In the United Kingdom, this was due largely to the sociopolitical atmosphere, which was decisively more favorable than those of other nations toward private enterprise.

Adam Smith was the primary interpreter of this extraordinary historical trend. He explained the particular economic fervor in England at the start of its industrial growth by contending that all men were naturally given to commerce and industry. He assigned to the state the limited but fundamental roles of defense, justice, and, in the economic arena, the creation of an infrastructure. France above all, and in a second wave Germany, put the burden of industrialization on the government. For the nation-state, economic development and political power went hand in hand, so either liberalism or protectionism could be applied, depending on their appropriateness. These practices are still in use today.

By the middle of the 19th century, the group of European nations that would lead the world economy had been defined. In the 1860s, decisive developments occurred in the United States with the Civil War, in Japan with the modernization projects of the Meiji, and in Russia with the abolition of serfdom. The picture of the future industrial powers was thus completed and would stand unaltered for over a century.

1. The Storming of the Bastille, *Musée Carnavalet, Paris.* On July 14, 1789, the people of Paris attacked the Bastille, an important political prison and thus a symbol of absolute monarchy. Following the convocation of the Estates General and the proclamation of the National Assembly, the direct actions of the populace forced the emergence of the social dimension of the revolution.

2. Napoleon on horseback. He became the symbol of French nationalism and of the struggles among nations.

3. Francisco Goya, The 3d of May, 1808: The Executions on Príncipe Pío Hill, *the Prado, Madrid.* The Napoleonic conquests fed the nationalism of European peoples. Particularly tenacious was the partisan warfare in Spain. In two celebrated paintings, Goya immortalized the beginning of the Spanish rebellion and the repression of the French troops.

4. Eugène Delacroix, Liberty Leading the People, *the Louvre, Paris.* Delacroix's work, considered the first political depiction in the history of modern painting, was inspired by the Parisian insurrection of July of 1830, the "Three Glorious Days". This uprising brought to power Louis Philippe, Duke of Orléans, a representative of the rich bourgeoisie.

5. Simón Bolívar, "the Liberator" (1783–1830). Although he was the main actor in Latin-American independence from Spain, Bolívar failed in his attempt to create a united South America. The examples of the United States and the French Revolution fed an anti-colonial fight that ended with the establishment of a series of republican countries that were far from democratic and marked by deep social inequality.

6. Benjamin West, Penn's Treaty with the Indians, *Pennsylvania Academy of the Fine Arts, Philadelphia.* William Penn (1644–1718), a major exponent of the Quaker movement, founded the Pennsylvania colony in 1682, providing it with a constitution of tolerance based on friendly relations with the American Indians. The colonies of New England were established by members of religious minorities who had emigrated from Europe and had settled, often hastily, on lands that would become the United States.

8. Assault and destruction of the tea ships at the Boston Tea Party on December 16, 1773 (in a print from that period). The fight for independence hinged on the colonists' refusal to pay taxes on English products. The colonies organized an army under the command of George Washington and on July 4, 1776 proclaimed the Declaration of Independence, giving rise to the United States of America.

9. Diorama with a model of the Boston Tea Party, housed at the Boston Museum.

7. The title page of the original edition of Adam Smith's An Inquiry Into the Nature and Causes of the Wealth of Nations, *London, 1776.* The Wealth of Nations is considered the first organic treatise on political economics. According to Smith, the welfare of nations depends on the number of productive workers, the extent of the division of labor, and the expansion of markets; these bases, combined with the free rein of the "invisible hand" that regulates supply and demand, allow for theoretically limitless development.

SOCIAL AND POLITICAL MOVEMENTS

10. BROTHERS AT WAR: SOCIALISM AND NATIONALISM

Before differentiating itself from communism and anarchism, socialism arose as a criticism of capitalism, which was based on the private ownership of property. It advocated the common possession of goods and of the means of production. Its birth was directly tied to the Industrial Revolution, but its inspiration descended from the utopias and the movements for social regeneration that had been continuous throughout antiquity and the Middle Ages—and that continues into modern times.

The economic progress of the 18th and 19th centuries rewarded the entrepreneurial bourgeois, while large sectors of the working classes were victims of the repercussions of the factory system. This same happened to the weavers, who at one time had worked at home on handlooms but now were forced to labor under extremely difficult conditions —as often was also the case for women and children.

These conditions undoubtedly were the springboard for the clashes of the time and for the first labor organizations, and they were the hinterland in which socialist and utopian ideologies matured. Based on the ideas of Owen, Fourier, Blanc, Saint-Simon, and Cabet, cooperative and associative experiments multiplied. Their goal was to create a new society, parallel or alternative to the one that existed at the time.

On the other hand, Marx and Engels claimed that in order for communism to be realized, political power would have to be seized by the working classes. They saw communism as the full realization of human potential—beyond all the traditional divisions of class, nation, race, and religion.

Even though the French Revolution brought about the establishment of the French nation-state, the modern political force of nationalism was born in opposition to it. The European nations, through Romanticism, affirmed their own historic and cultural individuality against the leveling force of revolutionary and Napoleonic France. Romanticism provided political nationalism with an elaboration of the concept of "nation." The nation thus embraced all of the existence of a people; according to Alessandro Manzoni's well-known definition, the nation became "one in weapons, one in language, one in altars, one in memories, one in blood, one in heart."

Nationalism revealed itself as a deep and pervasive current in modern life. It was present to various degrees within all the other ideological families, from liberal to socialist, from democratic to racist. Nationalism could feed imperialism as well as the anti-colonial struggle, 20th-century totalitarianism as well as the resistance that arose against it. It offered a remedy for the perceived social dissolution brought on by industrial capitalism. Socialism and

1. Johann Gottlieb Fichte (1762–1814) in a portrait by E. Gebauer. An idealistic philosopher who incited resistance against the French in his Addresses to the German Nation *(1807–1808), Fichte was the inspiration for the nationalistic myth of German superiority over other people. According to Fichte, the nation was something original and eternal, divine. He thus established the basis for nationalism as a political religion.*

2. François Perrin's Giuseppe Mazzini, *1851. Mazzini, who was born in Genoa, Italy in 1805 and died in Pisa in 1872, was one of the principal inspirations of the Risorgimento. His political ideas, heavily influenced by Romantic ideals, were centered on the concept of nationhood. Mazzini's mission, carried out with a tireless spirit of sacrifice, was to emancipate the Italian nation. According to Mazzini, the nation was the meeting point between people and intellectuals, the way to overcome the antagonism that resulted from class differences.*

3. In this depiction (France, 1848), a worker in the ateliers nationaux ["ah-tul-yay nah-syuh-noh," national workhouses] speaks with a bourgeois, a member of the middle class. Following the February Revolution, Louis Blanc established the ateliers nationaux to ensure work for the unemployed. The abolition of the workshops was one of the causes of the workers' insurrections of June of 1848. The uprisings were repressed with much bloodshed. The entire period of France's Second Republic (1848–1852) was marked by conflict between the proletariat and the bourgeoisie.

4. Gustave Courbet, Portrait of P.J. Proudhon in 1853, *the Louvre, Paris. Pierre-Joseph Proudhon (1809–1865) was one of the most important ideologues of the workers' movements. A fierce adversary of Marx, he called for a reconciliation of socialism and a market economy; he favored mutualism and federalism and opposed collectivism and statism. Proudhon's individualistic positions found favor with the large 19th-century artisan population.*

5. Karl Marx (1818–1883) in a photograph taken in 1867. Marx was the main theorist of modern communism. He outlined his program with Friedrich Engels in their 1848 The Communist Manifesto. *He dedicated himself above all to the criticism of the political economy and to the study of industrial capitalism in relation to historical materialism.*

6. The first edition of the book Das Kapital [Capital], *Hamburg, 1867. Marx's colossal work, which was never finished and was published in part after his death, aimed to study capitalism scientifically—to identify its workings, its strengths, and its irresolvable contradictions on the levels of great industry and the world market.*

nationalism traditionally embodied the political division between the Left and the Right, and both opposed economic liberalism.

Still, in the concrete historical context curious interlacings can be seen. This was because all these political and economic ideologies met on a kind of middle ground—a faith in industrial development—and thus found their affirmation within the state.

ATTITUDES AND CULTURES
11. FOR AND AGAINST INDUSTRIAL MODERNISM

It was only in the second half of the 19th century that the expression *Industrial Revolution* came into common use among historians. It is doubtful whether contemporaries of the great transformation that had hit England beginning in the last decades of the 18th century perceived the significance of the changes that were taking place.

To many at the time, the movement confirmed the idea that an unstoppable process governed history—that an "invisible hand" guided the development of production and consolidated individual interests, leading to improvements that progressively extended to all of society. Even among those who understood the tensions caused by the changes that were taking place, there were some who believed that the transition between the old agrarian society and the new industrial one could occur without traumatic disruptions.

Optimism was expounded by the French followers of the Enlightenment movement like Condorcet; by English philosophers and economists like Jeremy Bentham and Adam Smith; by the social experiments of utopianists like Robert Owen; and by the theories of Saint-Simon, champion of the belief in the capacity of industrialism to produce social justice. But other, profoundly different points of view were also being presented. At the threshold of the 19th century, Malthus posited that the population, which was increasing at a more rapid pace than was the availability of foodstuffs, posed a threat that needed urgently to be addressed. The degradation of the English cities of Manchester and Birmingham, even of London, was described in the inquiries of Engels and the novels of Dickens. Mary Shelley explored in *Frankenstein* the danger involved when science oversteps the boundaries of human potential.

The form of energy that was the basis for most of the innovations that were appearing caused enthusiasm as well as apprehension—after all, explosions of steam furnaces aboard ships and locomotives were not all that rare.

In his *The Philosophy of Manufactures,* Andrew Ure considered the period's division of labor and increasing use of machinery substantial and undeniable progress; but Karl Marx considered the same aspects of industrialization to be the source of a radical loss of the sense of work and, more generally, of human identity.

Even those—like Carlyle and Coleridge—who upheld the hopeful belief not in a movement beyond industrial capitalism to an egalitarian society, but in a return to a traditional society, harshly criticized the aridity of social relationships that were dominated by the ideas of economic competition and of production.

While the cultural world was divided in its judgement of the outcomes of the transformations that were taking place, the working classes were becoming aware of a profound change in the organization of daily life.

The punctuality necessitated by factory work hours, and the intense and monotonous rhythms provided by the machines, imposed an idea of time that was no longer elastic, measured by the task being performed, but rigid and uniform. In fact, the idea of being constantly oriented toward the future was spreading from the factory throughout society. The longstanding belief among merchants that "time is money" had not only overtaken the new industrial entrepreneurs but was becoming a widely diffused conviction. The belief was no longer in passing time, but in spending it.

In Lewis Mumford's words, "The clock, not the steam engine, was the most important machine of the modern era."

1. The clock controlled the exact hour, and thus the regularity of the industrial work hours and of the work day, based on the entrance and exit from the factory. The reorganization of the social conception of time was a key effect of the diffusion of the factory system.

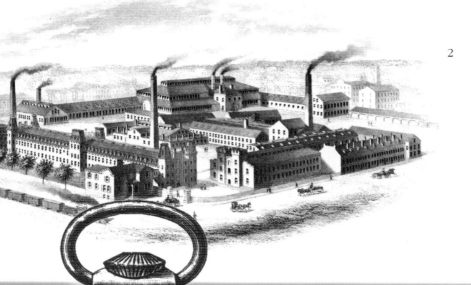

2. A later 19th-century American complex for the production of cotton. In the 19th century, industrial iconography often offered the idealized image of industrial centers in which factories merged harmoniously with the landscape and production with daily life.
3. Bell Time, an engraving by Winslow Homer from 1868, represents the departure of workers from a factory in Massachusetts.

4. and 5. Critical opinion on industrialization in the 19th century focused on recurrent images: the poverty of much of the urban population and the resultant degradation—in this case, the new production's pollution of the River Thames.

INDEX

Each entry is followed by the number(s) of the chapter(s) in which it appears.

Africa 1
agricultural revolution 1, 4
agricultural machines 1, 5
agriculture 1
air circulators 6
Alps (near Bergamo, Italy) 6
America 4
American Declaration of Independence (1776) 9
American Civil War 9
anarchism 10
Andersen, Hans Christian 7
anti-colonialism 9, 10
Arkwright 4
Ateliers nationaux (national workshops) 10
Atlantic 8

Bastille 9
Bentham, Jeremy 11
Birmingham 11
Blanc, Louis 10
blowing machines 5
Bolívar, Simón 9
Boston Tea Party 9
Bourgeois, entrepreneurial 10
Bourne, J. C. 1
brewers 5

Cabet, Étienne 10
Carlyle, Thomas 11
cast iron 5
charcoal 5
child labor 4, 6
children 4, 6, 8
city(ies) 3
Civil War, American 9
clock 11
coal 5, 6, 7
coke 5
Coleridge, Samuel Taylor 11
collectivism 10
colonies 4
colonists 9
communications 6
communism 10
company town 3
Condorcet, Antoine Nicolas de 11
Constitution of the United States of America (1787) 9
continuous-paper machines 8
corn 2
cotton mill 4, 11
cotton 4
Courbet, Gustave 10
Cromford 4

D'Alembert, Jean-Baptiste 1
daily life 7
Darby, Abraham 5

Davy, Humphrey 6
Degas, Edgar 3
Declaration of Independence (1776) 9
Delacroix, Eugène 9
demographics 2
Dickens, Charles 11
Diderot, Denis 1
discipline 4
distillation 5
division of labor 4, 9, 11
Doré, Gustave 3, 7

elevator hoists 6
energy 5, 11
Engels, Friedrich 10, 11
England 4, 5, 8, 9, 11
English Channel 8
Entrepreneurial bourgeois 10
entrepreneurs 11
environmental conditions 4
environments 3, 6
epidemics 2
establishment 4
Europe 3, 4, 8
explosives 6

factories 3, 4, 5, 7, 11
factory system 4, 10, 11
famines 1
farmers 3
federalism 10
Fichte, Johann Gottlieb 10
film 7
firedamp 6
foodstuffs 11
forests 5
Fourier, Charles 10
France 1, 5, 7, 8, 9, 10
Frankenstein 11
French Revolution 8, 9, 10

gas: introduction, 5, 6
General States 9
Genoa, Italy 10
Germany 2, 3
Goya, Francisco 9
Great Eastern 8

Hamburg 10
handloom 10
Heine, Heinrich 7
Hoe, Richard 8
hoisting machines 6
Homer, Winslow 11

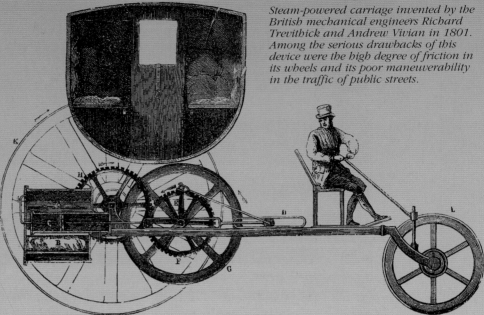

Steam-powered carriage invented by the British mechanical engineers Richard Trevithick and Andrew Vivian in 1801. Among the serious drawbacks of this device were the high degree of friction in its wheels and its poor maneuverability in the traffic of public streets.

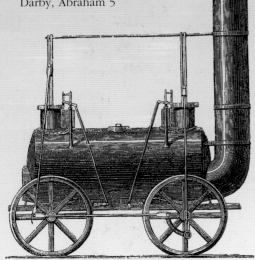

Left: Locomotive built by George Stephenson in 1815.